Elisabeth Regina Alves C. Silva
José Gustavo da Silva Melo
Luiz Carlos da Silva

Vegetation indices for analysing environmental degradation

Elisabeth Regina Alves C. Silva
José Gustavo da Silva Melo
Luiz Carlos da Silva

Vegetation indices for analysing environmental degradation

Case study on the Paratibe River, Paulista-PE

ScienciaScripts

Imprint

Any brand names and product names mentioned in this book are subject to trademark, brand or patent protection and are trademarks or registered trademarks of their respective holders. The use of brand names, product names, common names, trade names, product descriptions etc. even without a particular marking in this work is in no way to be construed to mean that such names may be regarded as unrestricted in respect of trademark and brand protection legislation and could thus be used by anyone.

Cover image: www.ingimage.com

This book is a translation from the original published under ISBN 978-613-9-64874-0.

Publisher:
Sciencia Scripts
is a trademark of
Dodo Books Indian Ocean Ltd. and OmniScriptum S.R.L publishing group

120 High Road, East Finchley, London, N2 9ED, United Kingdom
Str. Armeneasca 28/1, office 1, Chisinau MD-2012, Republic of Moldova, Europe
Printed at: see last page
ISBN: 978-620-7-79872-8

SUMMARY

Due to the intense urban growth of cities, green areas and water bodies are being pressurised by the environmental changes that have taken place, mainly as a result of urban growth. In this sense, there has been growing concern about ways to control and mitigate environmental degradation, which requires prior planning of land use. A survey of current land use, which is necessary for planning purposes, can be obtained using multispectral data provided by remote sensing satellites. In this way, the case study addressed in this work takes place in the area of influence of the Paratibe River, in the state of Pernambuco, and aims to carry out a spatio-temporal analysis of the lagoon based on products obtained through remote sensing. Three biophysical variables were used in the evaluation: the normalised difference vegetation index (NDVI) to assess the chlorophyll content of the vegetation, the soil-adjusted vegetation index (SAVI) and the leaf area index (LAI) to understand the dynamics of the area. Two Landsat 5 satellite images from 2005 and 2011 were used. The study showed that the vegetation with considerable chlorophyll content decreased considerably over two decades in the areas around Mata do Janga, but within the reserve the vegetation increased between 2005 and 2011, due to the area's conservation strategies. SAVI and IAF, on the other hand, decreased during the sampling period as they are parameters linked to vegetation structure.

Keywords: vegetation indices, urban densification, vegetation structure, conservation areas.

SUMMARY

CHAPTER 1

INTRODUCTION

The Atlantic Rainforest is found all along the Brazilian coast, both in the coastal region and on the plateaus and mountain ranges of the interior (NEVES, 2006). The Atlantic Rainforest is one of the two largest and most important tropical forests on the South American continent. Originally it covered more than 15% of the current Brazilian territory, but today, due to the various economic cycles and the expansion of urbanisation, the total area of this forest has decreased (RESERVA DA BIOSFERA DA MATA ATLÂNTICA, 2003).

Before its intense destruction, the Atlantic Forest covered large areas along the Brazilian coast, from the north-east to the south of the country, moving inland to varying extents, with a great diversity of soils, reliefs and climates, having as a common element the exposure to the humid winds that blow from the ocean (CORRÊA, 1995).

According to data from the SOS Mata Atlântica Foundation, its original area continuously covered an area of 1,363,000 km^2 , which corresponded to around 16 per cent of Brazil's territory. It covered all or part of 17 states in the country: Piauí, Ceará, Rio Grande do Norte, Paraíba, Pernambuco, Alagoas, Sergipe, Bahia, Espírito Santo, Minas Gerais, Rio de Janeiro, São Paulo, Paraná, Santa Catarina, Rio Grande do Sul, Goiás and Mato Grosso do Sul.

Despite the enormous lack of specific data proving the process of devastation, it is widely known that the dense and diverse forest found by the colonisers has, over the centuries, been intensively exploited and/or replaced, and is currently considered to be one of the most threatened ecosystems in the world (CORRÊA, 1995).

Data from Gonzaga de Campos (1912) indicated that the Northeast had 36.8% of its area covered with forest at the beginning of the century and, for the state of Pernambuco, 34.14% at the time. The survey of the Atlantic Forest of Pernambuco, carried out in 1993 (Braga et al, 1993), indicated a percentage of approximately 4.6% in relation to the original area, including the remaining areas of the associated ecosystems (mangroves, restingas and highland swamps) and only 1.5% remaining in relation to the area of the state (COSTA-LIMA, 1998).

The Atlantic Rainforest is currently Brazil's most threatened forest, with only 12.5 per cent of its original area preserved. Because it is so threatened and still harbours a high number of endemic species, it has been classified as one of the world's 25 biodiversity hotspots (MYERS et al., 2000; BRAZ et al 2011). In the state of Pernambuco, the Atlantic Forest has been reduced to 4.6% of its original cover (COSTA-LIMA, 1998). The capital of Pernambuco is 100 per cent part of the biome and has 20 per cent of it preserved, a percentage that corresponds to 4,400 hectares (Fundação SOS Mata Atlântica, 2015; INPE, 2015).

It is precisely in coastal areas that the Atlantic Rainforest suffers the greatest impacts. Property speculation, demographic pressure and unregulated occupation stimulate environmental degradation. This can be seen in the municipality, where the beaches are very popular with tourists and large property developments are built, further accelerating degradation. The accelerated and poorly planned growth of cities is causing serious damage to nature, including the deforestation of native vegetation to structure the urban network and all its infrastructure, atmospheric, water and soil pollution, and especially the destruction of various ecosystems, also accelerating urban problems and disputes over access to housing and infrastructure (SILVA et al, 2008).

The vegetation cover of the North Coast of Pernambuco, in its original composition, corresponds to the Atlantic Forest, whose exuberant vegetation and biological diversity have been destroyed since colonial times by the cultivation of sugar cane and coconut. Nowadays, this destruction has been carried out by the establishment of allotments for farms and recreational farms and by the extraction of wood and firewood for consumption in urban and rural areas. The most extensive remnants of forest on the North Coast are located to the west of the BR-101 motorway and the urban centres that border it. They are concentrated in the municipalities of Abreu e Lima, Igarassu, Itamaracá, Paulista and southeast of Itaquitinga, while they are sparse in the rest of the area (CPRH, 2003).

In Paulista, Abreu e Lima and Igarassu, these remnants generally occur in the western part of the valleys of the Paratibe, Barro Branco, Utinga, Bonança, Tabatinga, Arataca and Botafogo rivers, as well as on the Araçá and Chã da Cruz plateaus. Patches of vegetation that are being recomposed sometimes connect these remnants. In the eastern part of these municipalities, the forests are located inside the urban area or close to it, associated in some cases with capoeiras of different sizes (CPRH, 2003).

Given this heterogeneous scenario, remote sensing is an alternative for diagnosing the state of the landscape in relation to a degraded ecosystem such as the Paratibe River in Paulista-PE. Quantification, risk assessment and environmental monitoring can be carried out using biophysical parameters (such as vegetation indices) and physical parameters (albedo, temperature, emissivity, among others) obtained from orbital images to determine changes in the surface (LOPES, 2010).

In view of this, vegetation indices can provide an overview of the situation in areas with mangrove vegetation, as they are some of the most widely used parameters in the seasonal and inter-annual monitoring of physiological and structural parameters of different ecosystems and are obtained through the use of remote sensing. They consist of linear or non-linear transformations of spectral bands chosen according to their specific characteristics in order to emphasise the contribution of vegetation properties of interest (SOUZA, et al., 2009).

Using remote sensing, it is possible to analyse, among other variables, the chlorophyll content present in the vegetation through the normalised difference vegetation index (NDVI), the areas of degradation present in the perimeter studied through the soil-adjusted vegetation index (SAVI) and the leaf area index of the area

(IAF), (SILVA et al, 2015).

Vegetation indices such as the normalised difference vegetation index (NDVI) can therefore help to monitor certain biomes, as well as making it possible to analyse the natural and man-made landscapes of their various ecosystems, thus improving understanding of their structure, functioning and ecological function (LOPES, 2010).

Since the 1960s, the study of the spectral characteristics of vegetation, based on parameters such as biomass and Leaf Area Index (LAI), mainly in the visible and near infrared bands, has been increasingly developed due to the important results found in its applications, which provide data for edaphic, climatic, temporal and phenological analyses. The leaf area index (LAI) defined by Watson (1947) as the integrated leaf area of the canopy per unit area projected onto the ground (m /m^{22}) is one of the main biophysical variables of a forest canopy and is directly related to the productivity and evapotranspiration of forest ecosystems, considering the surface area of just one of the leaf faces in the calculation (LANG and MCMURTRIE, 1992).

According to Clough (1992), IAF has a major influence on the primary production and growth of mangrove plants and is therefore useful for analysing the phenological stage of plants. In addition to this analysis, climatic factors such as solar radiation, day length, air and soil temperature, rainfall and potential evapotranspiration, as well as their seasonal variability, are essential in damaging the growth of mangrove forests around the world. This vegetation index is important for quantifying the level of biomass variation over the years, given that the presence of vegetation is intrinsically correlated with the presence of moisture in the air and thermal comfort, playing a fundamental role in regulating the microclimate.

In this way, the physiognomic characteristics of vegetation play a fundamental role in microclimate studies of heterogeneous environments such as the Paratibe Basin in Paulista-PE. For this reason, the aim of this work is to carry out a spatio-temporal analysis of the area's vegetation dynamics using Vegetation Indices, identifying the green areas on the banks of the river, their vegetative vigour and the structural aspects of the vegetation.

CHAPTER 2

OBJECTIVES

2.1 General

Spatio-temporal analysis of environmental degradation on the banks of the Paratibe River in the state of Pernambuco using vegetation indices.

2.2 Specific

- ❖ Estimation of the *Normalised Difference Vegetation Index* (NDVI);
- ❖ Estimation of the vegetation index Leaf Area Index (SAVI);
- ❖ Analysis of the geographical systems that make up the Paratibe River in Paulista-PE;
- ❖ Analysis of the biophysical components that characterise the Paratibe River in Paulista, PE.

CHAPTER 3

THEORETICAL FRAMEWORK

3.1 Characterisation of the Paratibe River Basin

The study of river basins makes it possible to integrate the factors that condition the quality and availability of water resources with their real physical and anthropic conditioning factors (Valente, 1976; Hein, 2000; Grossi, 2003; Prado, 2005).

According to Moldan and Cerny (1994), from a hydrological point of view, the microwatershed can be considered the smallest landscape unit capable of integrating all the components related to water quality and availability, such as the atmosphere, natural vegetation, cultivated plants, soils, underlying rocks, bodies of water and the surrounding landscape. Environmentally, it can be said that the watershed is the ecosystem and morphological unit that best reflects the impacts of anthropogenic interference, such as land occupation with agricultural activities (JENKINS et. al., 1994). Lima (1999) comments that the microwatershed is the well-defined manifestation of an open natural system, which can be seen as the ecosystem unit of the landscape.

It is in the river basin that most of the consequences or impacts of the use of natural resources manifest themselves. They manifest themselves as: soil erosion, nutrient leaching, sedimentation of rivers, lakes and reservoirs, degradation of riparian forests and vegetation cover, deposition of solid waste, waste and effluents resulting from human activities, floods and inundations, water-borne endemics and epidemics. This is one of the reasons why the basin is taken as the physiographic planning unit. Given the legal provisions inherent to irrigation, water resource management and the environment, any irrigation and drainage project, whether collective (community of irrigators) or individual, must refer to the river basin in which it is located, at least in order to comply with the conditions for granting and, where appropriate, charging for the use of water resources (MINISTÉRIO DA INTEGRAÇÃO NACIONAL, 2008).

According to the Socio-Environmental Diagnosis of the North Coast (GERCO, 2003), the municipality of Paulista is located in the Paratibe and Timbó river basins, in coastal micro-basins and in a small part of the Igarassu river basin. The percentage of the municipality's area in each basin corresponds to: the Igarassu river basin (0.4%); the coastal micro-basins (10.1%); the Timbó river basin (25.7%) and the Paratibe river basin (63.8%) (GERCO, 2003).

The Paratibe River basin is the largest of all and covers around 11,800 hectares. It covers land in the municipalities of Paulista, Olinda, Recife and Camaragibe. The part that falls within the municipality of Paulista corresponds to 6,283.09 hectares. It is bordered to the north by the Timbó and Igarassu river basins; to the south by the Beberibe river basin; to the west by the Capibaribe river basin; and to the east by the micro-basins draining the marine terrace (GERCO, 2003).

The most common forest formations present in the area are:

- OMBRÓFILA FOREST - Native forest formation in a medium to advanced stage of regeneration associated with remnants of Atlantic forest and Restinga forest. Tree physiognomy, with a closed canopy, the presence of epiphytes and a diversified understorey. In the IDA, this formation was identified in the fragments upstream of the BR-101 motorway, in the Janga and Jaguarana forests and in the interfluve of the Paratibe River and the Tintas Canal in the final stretch near the PE-001 motorway. These are arboreal vegetation bodies of the Atlantic Forest, with tall trees. The predominant phytophysiognomy of these forest fragments is classified as open Ombrophilous Forest, depending on the degree of anthropogenic pressure. Smaller, more anthropised fragments are commonly referred to as capoeiras. These formations were identified in the Matas de Jaguarana in Paulista Centro, in the Mata de Maranguape, upstream of the BR-101 and at the old airfield near PE-022. These fragments have been isolated by urban agglomerations and suffer from anthropogenic pressure in their surroundings.

- Coastal Paludosa Natural Forest Formation - MANGUE. Vegetation with fluvial-marine influence, identified in Section 03, covering part of the lower lands of the Paratibe River interfluve and the Tintas Canal.

- MIXED FOREST FORMATION made up of patches of fruit trees and native species, with densities ranging from open to medium. These areas are inserted between herbaceous fields and/or urban areas. Mixed physiognomy that is sometimes shrubby with the presence of native and exotic species and some areas with herbaceous cover, sometimes dominated by a herbaceous matrix and small fragments of fruit trees and native species scattered in a very sparse or more dense manner.

3.2 Definition of Atlantic Forest

Because they are located in the tropics, tropical forests are known for their high biodiversity, which is one of the factors that contributes to the existence of many ecological niches (MANTOVANI, 2003). In Brazil, there are two important large tropical forests: the Amazon Rainforest and the Atlantic Rainforest.

The Atlantic Rainforest owes its existence to climatic types ranging from hot and humid to moderately cold (mesothermal), which prevail on Brazil's Atlantic coast. High temperatures, high relative humidity, abundant rainfall, frequent fog in some areas and intense light characterise these climates. The diversity of the terrain contributes regionally to the structural changes in the forest (PEREIRA, 2009).

Originally, this vegetation formation stretched along Brazil's Atlantic coastline, from the Trairi River in the south of Rio Grande do Norte to the Tapes and Herval mountains in Rio Grande do Sul, to the west of Lagoa dos Patos. Narrow in the north-east, it widened southwards until it reached its maximum width in the Paraná River basin, even penetrating Paraguayan and Argentinian territory. Successive episodes of devastation have reduced its original area, which currently stands at around 6 to 8 per cent of its original area (PEREIRA, 2009).

According to data released by the SOS Mata Atlântica Foundation and the National Institute for Space Research (INPE) in the "Atlas of Atlantic Forest Municipalities" (2015), the capitals of Rio Grande do Sul and Santa Catarina are in first and second place in the Atlantic Forest preservation ranking, with 32 per cent and 25 per cent of preserved Atlantic Forest, respectively. Among the capitals of Brazil, Recife is the third most preserved Atlantic Forest, proportionally. The capital of Pernambuco is 100 per cent part of the biome and has 20 per cent of it preserved, a percentage that corresponds to 4,400 hectares. In Pernambuco, the municipality of Goiana, Mata Norte, tops the list of municipalities with the most deforestation between 2000 and 2014. It deforested 86 hectares of Atlantic Forest during this period. The municipality of Abreu e Lima stood out in terms of preservation. The region managed to maintain 62.1 per cent of its natural vegetation.

The first specific legal norm for the Atlantic Forest was Decree 99.547/1990, which prohibited any suppression of native vegetation in the Atlantic Forest. This was replaced in 1993 by Decree 750, which legally defined the domain, including different forest formations and associated ecosystems, and determined the protection of the remaining primary native vegetation, as well as secondary vegetation in regeneration. With the same guidelines as this decree, a draft Atlantic Forest Law was presented in 1992 by the then federal deputy Fábio Feldmann (MMA, 2010).

Law No. 11.428 of 2006, regulated by Decree No. 6.660 of 2008, provides for the conservation, protection, regeneration and utilisation of native vegetation, both forest formations and the associated ecosystems that make up the Atlantic Forest. In addition to Law No. 11.428 and Decree No. 6.660, there are Conama resolutions that define primary and secondary vegetation and provide the parameters for identifying primary vegetation and secondary vegetation in the initial, medium and advanced stages of regeneration, both in forest formations and in the vegetation of restingas and altitude fields (MMA, 2010).

In the Northeast, north of the São Francisco River, botanists and geographers have long recognised the existence of two different patterns of tropical forest. Referring to the Atlantic Rainforest, Vasconcelos Sobrinho (1970, p. 61) stated that:

> "...studying its presence in the Northeast, we found, as early as 1941, two marked differentiations, which we called the Humid Forest and the Dry Forest, diversified by a few species that are peculiar to them, but mainly because the Humid Forest is characteristically evergreen, while the Dry Forest is semi-deciduous, with the trees losing almost all their leaves in the dry season, as well as being taller and smaller in diameter than in the Humid Forest."

In addition to its great territorial extension, other geographical factors, such as the variation in altitudes, differences in soil and landforms, among others, provide the Atlantic Rainforest with extremely varied scenarios. For this reason, its domain is made up of various formations, such as dense ombrophilous, mixed ombrophilous, open ombrophilous, semideciduous seasonal, deciduous seasonal and high altitude grasslands, as well as associated ecosystems such as mangroves, restingas and inland swamps. According to Campanili and Prochnow (2006), these are the plant formations found in the Atlantic Forest Domain:

- Dense ombrophilous forest - Evergreen forest with a canopy of up to 15 metres, with emergent trees up to 40 metres high. Dense shrub vegetation, made up of arborescent ferns, bromeliads and palm trees. Climbers and epiphytes (bromeliads, orchids), cacti and ferns are also very abundant. In the wettest areas, sometimes temporarily waterlogged, there were fig trees, jerivás and palm trees (Euterpe edulis) before human degradation. It extends from Ceará to Rio Grande do Sul, located mainly on the slopes of the Serra do Mar, the Serra Geral and on islands on the coast between the states of Paraná and Rio de Janeiro.

- Mixed ombrophilous forest - Known as the Araucaria Forest because the Paraná pine (Araucaria angustifolia) forms the upper level of the forest, with a very dense understory. Before anthropogenic interference, this formation occurred in regions with a subtropical climate, mainly on the plateaus of Rio Grande do Sul, Santa Catarina and Paraná, and in discontinuous massifs in the higher parts of São Paulo, Rio de Janeiro and southern Minas Gerais (the Paranapiacaba, Mantiqueira and Bocaina mountains).

- Open ombrophilous forest - This is considered a transitional type of dense ombrophilous forest, occurring in environments with drier climatic characteristics. It is found, for example, in Bahia, Espírito Santo and Alagoas. Seasonal forest (deciduous and semi-deciduous) - Forest with trees between 25 and 30 metres tall, with the presence of deciduous species (they shed their leaves during the colder, drier winter), with a considerable occurrence of epiphytes and ferns in the wetter areas and a large number of lianas (vines). Before human degradation, they occurred to the west of the ombrophilous forests of the Atlantic slope, entering the Brazilian Plateau as far as the banks of the Paraná River. The Morro do Diabo State Park and the Iguaçu National Park protect this type of forest.

- Brejos interioranos - These are areas with a different climate in the interior of the semi-arid region, also known regionally as "humid mountains", as they originally occupied most of the tablelands and eastern slopes of the Northeast.

- High altitude grasslands - These occur at elevations above 1,800 metres and on localised ridge lines. The characteristic vegetation is made up of grass communities, in some places interrupted by small heathlands. Often at higher altitudes there are flat tops or rocky peaks, as in the Itatiaia National Park (located between Rio de Janeiro, São Paulo and Minas Gerais).

- Mangroves - Formation that occurs along estuaries, due to the brackish water produced by the meeting of fresh river water with the sea. It is a very characteristic vegetation, as it has only seven species of trees - less than 1% of those recorded in the Atlantic Rainforest - but is home to a diversity of microalgae at least ten times greater. This invisible forest, according to researchers from the Federal Rural University of Pernambuco (UFRPE), is capable of occupying a single square centimetre of mangrove root with around 200,000 representatives.

- Restinga - Occupies large stretches of the coastline, on dunes and coastal plains. It begins near the beach, with grasses and undergrowth, and gradually becomes more varied and developed as it moves inland, and may also have marshes with dense aquatic vegetation. It is home to many cacti, orchids and bromeliads. Today, this formation is largely devastated by urbanisation.

In the more extensive remnants of forest on the North Coast, the balance of devastation is represented by a few remnants of forest which, in general, cover the slopes of tablelands and hills with a high gradient and, to a lesser extent, hills and gentle patterns in the eastern part of the area, where urban occupation and the subdivision of land for farms and ranches have led to the almost total destruction of the existing forests (CPRH, 2003).

3.3 Description of the Restinga Forest

The restinga ecosystem, commonly seen as a coastal plain located between the beach and the foothills of the mountains, is much more than a transitional environment between the beach and the hillside. The Brazilian restinga is a set of coastal ecosystems with floristically and physiognomically distinct communities, which colonise sandy soils of very varied origins, form a complex edaphic vegetation and occupy sites as diverse as beaches, dunes and associated depressions, sandy ridges, terraces and plains. They are recognised as having three primary phytophysiognomies: herbaceous/sub-shrub, shrub and tree (FALKENBERG, 1999).

In a resolution issued by CONAMA (1985), we see the restinga as a coastal sandy accumulation, generally elongated and parallel to the coastline, produced by the accumulation of sediments transported by the sea, where characteristic mixed plant associations are found, commonly known as restinga vegetation.

The restingas are distributed along the entire Brazilian coastline, approximately 5,000 km, from the east coast of Pará to the coast of Rio Grande do Sul, sometimes extending for kilometres westwards into the interior of Brazil, providing great environmental and biological diversity in these ecosystems (MONTEIRO, 2014).

According to Otto (2013), restinga fragments occupy almost 79 per cent of the Brazilian coast. Its main formations occur on the coasts of São Paulo, Rio de Janeiro, Espírito Santo and Bahia and are the most affected environments, as the search for a second holiday home is common. The location most chosen by holidaymakers is on or near the beach, i.e. in the Restinga environment.

3.4 Situation of Atlantic Forest degradation in Paulista

Between the 1960s and 1980s, the municipality of Paulista was one of the largest textile producers in South America. Residents of the area reported that during the textile boom, twelve boilers burned 80 lorry loads of wood from the Atlantic Forest every day. This progress, coupled with population growth, led to major environmental degradation, especially in the area where the city centre is today. Between the 1970s and 1990s, new areas of modern industry were created in the municipality, encouraged by the former Northeast Development Superintendence (SUDENE). Today, most of them have been deactivated, also causing major losses of biodiversity, since they were installed in remnants of the forest (MUNICIPAL INFORMATION OF PAULISTA, 2008).

The waterfront in the municipality of Paulista has been the target of serious erosion problems, associated with the implementation of erosion containment works that fix the coastline but do not recover the beach, and the disorderly urban occupation of the post-beach strip (CPRH, 2003). In recent decades, coastal erosion has become a problem of increasing magnitude on the beaches of the central coast of Pernambuco. The first references to coastal erosion in the state date back to 1914 (FINEP/UFPE 2009), and report the damage caused by the construction of a jetty located near the Isthmus of Olinda.

In the municipality of Paulista, the erosion process is relatively recent, having been recorded since the 1980s and possibly associated with the construction of the jetties on the Paratibe River (also known as the Rio Doce) (COSTA 2002, FISNER 2008). This process intensified in the 1990s, requiring the construction of a series of works to contain erosion. In their 2002 studies, Manso et al. (2006) pointed to a situation of moderate erosion on Maria Farinha beach, with a reduction in the width of the post-beach and damage to structures located in different sectors of the beach system. On Pau Amarelo and Janga beaches, the erosion process was more intense at the time, with the same impacts observed.

Since then, impacts related to the erosion process have been observed in various parts of the coastline, especially in urban areas. In response to erosion and in order to control the process or mitigate its effects, municipalities have opted to fix the coastline, so that along the study area there are numerous interventions, often made without a good understanding of coastal dynamics. These interventions profoundly modify the coastal environment, entail high costs and have results that are not always satisfactory, as well as often transferring erosion to adjacent beaches (FINEP/UFPE, 2009) and causing a loss of the scenic beauty of the coastline.

3 . 5 Configuration of the Mata do Janga Urban Forest Reserve

According to Carvalho (2011), the Paratibe River is given special attention in Paulista's Municipal Legislation. In the municipality's current Master Plan, the river is included in the Special Urban and Environmental Conservation Zone (ZECUA), from the BR-101 bridge to its mouth. The delimitation of this zoning includes permanent protection areas, defined in federal legislation and characterised by significant vegetation cover, one of which is the Mata do Janga.

Located in the south-eastern part of the municipality of São Paulo, its area covers 132.24 ha, corresponding to 1.36% of the municipality's total area. It is part of the Paratibe River basin, an important spring that rises to the west of the region and cuts through the neighbourhoods of Paratibe, Arthur Lundgren I, Maranguape I, Bairro do Nobre, later joining the Fragoso River to form the Doce River that separates the city of Paulista from Olinda. The last remnant of Restinga Forest in Pernambuco, its relief is flat to undulating and it has a vegetation cover of secondary forest with medium and large-sized individuals and an irregular shrub layer, both in density and size, which contributes to urban environmental quality (MONTEIRO, 2014).

Andrade Lima (1960) in his work recognised the restingas of Janga/Maranguape and Porto de Galinhas in Pernambuco, which at the time was the most affected by human activity. Classed as restinga forest, Mata do Janga was created by State Law No. 9,987 of 1987 and elevated to the category of Ecological Reserve by Law No. 9,989 of 13 January 1987, and is currently included in an area of public utility of social interest for the purposes of expropriation, according to municipal decree No. 094 of 30/12/1985. Today, by municipal law, it is called the Mata do Janga Urban Forest Reserve (Municipal Law No. 14.324\2001).

The Mata do Janga Urban Forest Reserve is a protector of the estuarine and riparian zone, protecting the banks of the Paratibe River from leaching, promoting soil retention, protecting it from erosion and siltation and increasing habitats, thus helping to preserve the biodiversity of fauna and flora (MONTEIRO, 2014).

4 .6 Conceptualisation of Remote Sensing and Vegetation Indices

Remote Sensing is defined in different ways by various authors, the most usual definition being that adopted by Avery and Berlin (1992) and Meneses (2001): "A technique for obtaining information about objects through data collected by instruments that are not in physical contact with the objects under investigation". For Jensen (2007), Remote Sensing (RS) can be defined as the art and science of obtaining information about objects without direct physical contact with the object.

The use of vegetation indices such as the Normalised Difference Vegetation Index (NDVI), Soil-Adjusted Vegetation Index (SAVI) and Leaf Area Index (LAI) makes it easier to obtain and model biophysical plant parameters such as leaf area, biomass and percentage of ground cover, especially in the infrared region of the electromagnetic spectrum, which can provide important information on plant evapotranspiration (JENSEN, 2009; EPIPHANIO et al., 1996).

Vegetation index modelling is based on the opposite behaviour of vegetation reflectance in the visible region, i.e. the higher the plant density, the lower the reflectance due to absorption of radiation by photosynthetic pigments and the higher the plant density, the higher the reflectance due to scattering in the different layers of the leaves (BORATTO E GOMIDE, 2013).

These indices are considered to be indicators of vegetation growth and vigour and can be used to diagnose various biophysical parameters with which they have high correlations, including leaf area index, biomass, percentage ground cover, photosynthetic activity and productivity (PONZONI, 2001).

These indices have been successfully used to monitor changes in vegetation on a continental, regional and global scale (BANNARI et al, 1995 cited by BARBOSA, 2006). This success is due to the differential reflectance of chlorophyll in the visible and infrared wavelengths (BARBOSA, 2006).

CHAPTER 4

MATERIALS AND METHODS

4.1 Characterisation of the Study Area

The Paratibe River Basin covers 11,800 hectares and includes land in the municipalities of Paulista, Olinda, Recife and Camaragibe. The part within the municipality of Paulista corresponds to 6,283.09 hectares. It is bordered to the north by the Timbó and Igarassu river basins, to the south by the Beberibe river basin, to the west by the Capibaribe river basin and to the east by the micro-basins draining the marine terrace (Figure 1). **Figure 1** - Location of the study area.

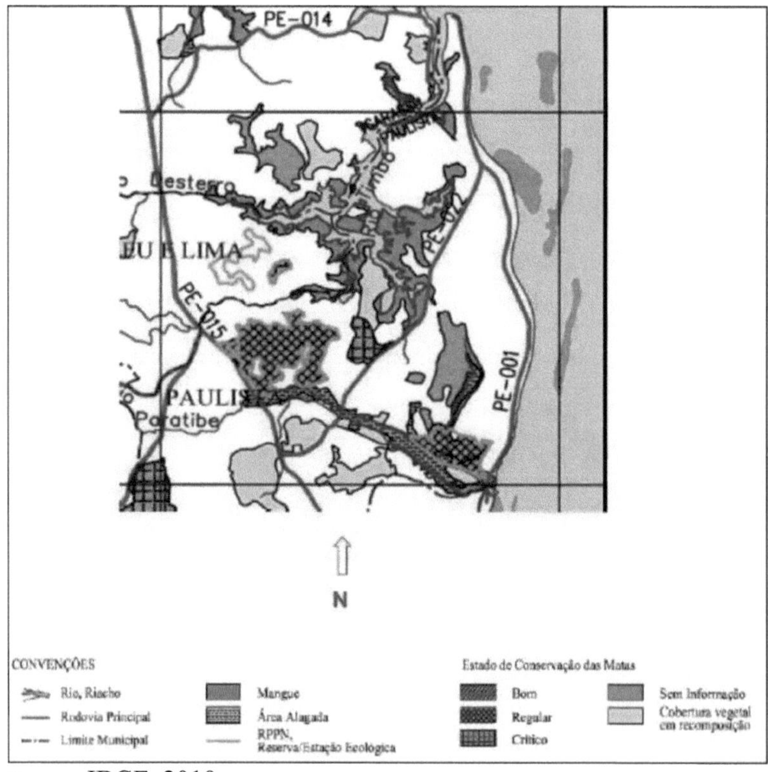

Source: IBGE, 2010.

According to the Socio-Environmental Diagnosis of the North Coast, the Municipality of Paulista is part of the Paratibe River and Timbó River Hydrographic

Basins, coastal micro-basins and a small part of the Igarassu River Hydrographic Basin (figure 1). The percentage of the municipality's area in each basin corresponds to:

- 0.4% in the Igarassu River Basin
- 10.1% in coastal micro-basins
- 25.7% in the Timbó River Basin
- 63.8% in the Paratibe River Basin

The Paratibe River rises on the border of the municipalities of Paudalho, Camaragibe and Paulista. Initially, it was called Riacho da Mina until it met the Riacho do Boi and was renamed Rio Paratibe. Its most extensive tributaries, on the right bank, are: the Canal das Tintas, the Fragoso and Piaba rivers and the Córrego Maximino; on the left bank, the Mumbeca River, the Riacho do Boi and the Riacho do Limoeiro flow into it (Figure 2). Due to the extensive occupation of its banks by isolated houses and COHAB housing estates, as well as by invasions located in the estuary itself, the area of the Paratibe River is not only relatively small, it is also quite de-characterised (Figure 2). These occupations have considerably reduced the area of mangroves, leaving only a small area at the mouth of the Paratibe and Doce rivers (Olinda).

Figure 2 - Route taken by the Paratibe and Doce Rivers

According to Pereira (2001), the dominance of phytoplankton along with the low values of specific diversity and number of species per season can be considered indicators of stressed environmental conditions, especially in the Rio Doce, which is directly influenced by domestic waste and the Paratibe River. Part of the wetlands stretching from the mangrove swamp to the western end of the PE-15 motorway, along the Paratibe River and the Canal das Tintas have been landfilled, as has the limoeiro stream to the west of the Janga neighbourhood. In order to check water pollution in the basins that drain into the Atlantic Ocean, Companhia Pernambucana do Meio Ambiente (CPRH, 2003) has been systematically monitoring these basins since 1984, and currently has a total of 196 sampling stations throughout the state.

The Paratibe River has three large protected areas: Mata do Janga, Mata dos Caetés and Mata do 7° RO. The main land uses are: Urban and industrial occupation, Atlantic Forest and Mangrove areas, and Polyculture and Forestry. Water use in the river is for public supply, domestic effluent reception and industrial effluent reception. The main industrial activities in the basin are: Food products, textiles, metallurgy, clothing/artifacts/textiles, paper/paperboard and mechanics.

The Janga neighbourhood is the most populous in Paulista, with around 44,008 inhabitants, according to the 2010 census. In this context, the beach not only receives a large number of tourists, but also people from other municipalities in the RMR, almost doubling its population during the summer season. As a result, property speculation in the neighbourhood began to increase, with the construction of condominiums and houses, increasing the neighbourhood's population.

Paulista is bordered to the north by the municipalities of Abreu e Lima and Igarassu; to the south by Olinda, Recife and Camaragibe; to the east by the Atlantic Ocean and to the west by the municipality of Paudalho. The municipality has an area of 101.80 km^2 , and a population of 300,466 inhabitants (IBGE, 2010).

The municipality of Paulista is part of the Metropolitan Development Region, according to the territorial division adopted by FIDEM (1987), which, according to the RMR Master Plan, is configured as an area that concentrates potential, understood as spaces or territories that propagate development or are strategic.

In Paulista, the remnants of the Atlantic Forest occur in the western part of the Paratibe river valley, known as Mata do Ronca. In the eastern part, the forests are located inside the urban area or close to it: Mata do Janga, Jaguarana, considered Urban Forests, recategorised by State Law 14.324/11 (AMANE, 2012) and the Caetés Ecological Reserve. These forests were created by Law No. 9,989 of 13 January 1987.

Of the three forests mentioned, only Caetés was established in 1991 and underwent a change of category, becoming an Ecological Station by State Law No. 11.622/98, mainly to help protect water resources, environmental education activities and scientific research, as well as providing leisure for the local population. The areas of the reserves are:

- Mata do Janga - 132.24 ha
- Jaguarana Forest - 332.28 ha
- Caetés Ecological Station - 150,00 ha

Figure 3 - Paratibe River (above) meeting the Fragoso River, forming the Doce River. Source: Google Earth, 2016).

Municipal Law No. 4040/2008 - Participatory Master Plan (replaced by Complementary Law No. 4253/2012), indicated the ZECUA - Special Urban and Environmental Conservation Zone in various parts of the municipality of Paulista and a specific sector called ZECUA Sector Via Parque do Paratibe, where it was intended to enhance the Paratibe River, figure 3.

Urban planning regulations have assigned ZECUAS the role of conserving their environmental amenity within the urban context, as well as containing the process of dispersed urban expansion, figure 4. Thus, the Paratibe ZECUA has the multiple function of protecting the watercourse and guaranteeing drainage by minimising the occurrence of flooding, but it also has an important function in urban terms and in terms of urban mobility (CPRH, 2013) figure 4.

Figure 4 - Paratibe ZECUA Via Parque sector.
Source: CPRH, 2013.

The table below shows the concentration of proposals focused on municipal and inter-municipal mobility (43 per cent), followed by economic development proposals (28 per cent). In general, the mobility-related proposals influence the project under analysis, as they are at higher levels of progress than the Paratibe ZECUA, either in the process of being implemented or at executive project level (Figure 4).In any case, it is clear that new roads need to be built and existing roads upgraded as part of the solution to a problem that has become chronic in the RMR, and which has set the 2014 World Cup as a first evaluation point, where it is hoped that substantial progress will be made in this field by then (Figure 5).

Figure 5 - Overview of the North Way complex in Paulista. Source: CPRH- RIMA, 2013.

Source: CPRH, 2013.

In the forests of the North Coast, you'll find, among other species, cupiúba (Tapirira guianensis), cabotâ-deleite (Thyrsodium schomburkianum), sucupira branca (Bowdichia virgiloides), laurel (Ocotea spp), embiriba (Eschwelera ovata), murici da mata (Byrsonima sericea), barbatimão (Abarema cochliocarpos), ingá (Inga spp), visgueiro (Parkia pendula), embaúba (Cecropia adenopus), cajueiro (Anacardium occidentale), paquevira (Heliconia angustifolia), pereira da mata (Luchea ochrophylla), pau d'arco (Tabebuia sp), camaçari (Caraipa densifolia), munguba (Bômbax gracilipes), embiridiba (Buchenavia capitata), (CPRH, 2003).

With regard to the fauna of the north coast of Pernambuco, the literature cites the occurrence of approximately 200 animal species, including mammals, birds, reptiles and amphibians. Information gathered from the local population during the field research revealed that some of these species occur in the area, such as: black anu

(Crotophaga ani), white anu (Guira guira), bem-te-vi (Pitangus sulphuratus), vulture (Coragyps atratus), sparrow (Passer domesphuratus), sagui (Callithrix jacchus), sloth (Bradypus variegatus), armadillo (Dasypus novemcintus), paca (Agouti paca), agouti (Dasyprocta sp), marmoset (Cavia aparea), chameleon (Iguana iguana) and marsh harrier (Ameiva ameiva).

Although some remnants of forest on the North Coast are likely wildlife refuges due to their size, deforestation and predatory hunting have led to a reduction in plant and animal species, thus threatening the floristic and faunal diversity that still exists in the area (CPRH, 2003).

In the coastal strip, on land subject to the constant influence of the tides, mangrove vegetation develops, with the most common species being the red mangrove (Rhizophora mangle), the white mangrove (Laguncularia racemosa) and the siriúba mangrove (Avicennia), as well as less frequent species such as the button mangrove (Conocarpus erectus), the mangrove fern (Acrostichum aureum), the reed (Eleocharis), the tiririca (Scleria bracteata), among others. In the marshy areas upstream of the mangrove, typical hygrophilous vegetation develops, often dominated by the aninga (Montrichardia linifera) (CPRH, 2003).

4.2 Satellite Image Processing

4.2.1 <u>Space-Time Analysis</u>

Initially, the study area was delimited. Images from 2005 and 2011 from the Landsat-5 satellite were interpreted and analysed, downloaded free of charge from the website of the National Institute for Space Research (INPE) and processed at the UFPE Remote Sensing and Geoprocessing Laboratory (SERGEO), which holds the licence for the software used in this work.

4.2.2 <u>Image processing and layout assembly</u>

Initially, all the images were registered in ERDAS Imagine 9.3 software using an image that had already been georeferenced. The images were downloaded from the

INPE website. To process the Landsat-5 satellite image, models were created using the Model Maker tool in the ERDAS Imagine 9.3 software under licence from the Department of Geographical Sciences at the Federal University of Pernambuco.

4.2.3 Radiometric calibration

The set of radiance or radiometric calibration is obtained using the equation proposed by Markham and Baker (1987) (Equation 1):

$$L\,\lambda i = \alpha t + \frac{bt - \alpha t}{255} ND \quad (1)$$

Where a and b are the minimum and maximum spectral radiances (1 1 2pm srWm), ND is the pixel intensity (integer between 0 and 255) and i corresponds to the bands (1, 2, ... and 7) of the Landsat 5 and 7 satellite. The calibration coefficients used for TM images are those proposed by Chander and Markham (2003) and Oliveira et al, 2010.

4.2.4 Reflectance

The reflectance of each band (i) is defined as the ratio between the solar radiation flux reflected by the surface and the incident global solar radiation flux, obtained using the equation (ALLEN et al., 2002 apud OLIVEIRA et al, 2010), (Equation 2):

$$\rho\lambda i = \frac{\pi\,.L\lambda i}{K\lambda i\,.\cos Z\,.dr} \quad (2)$$

Where λiL is the spectral radiance of each band, λik is the spectral solar irradiance of each band at the top of the atmosphere 12μm, Z is the solar zenith angle and rd is the square of the ratio between the mean Earth-Sun distance(ro) and the Earth-Sun distance(r) on a given day of the yearDSA),(OLIVEIRA et al, 2010; SILVA et al, 2011).

4.2.5 *Normalised difference vegetation index* (NDVI)

The Normalised Difference Vegetation Index (NDVI) is the ratio of the difference between the reflectivities of the near infrared and red bands and the sum of these reflectivities (ROUSE et al., 1973). NDVI is a sensitive indicator of the quantity and condition of vegetation, whose values range from -1 to 1. On surfaces containing water or clouds, this variation is always less than zero0).

The Normalised Difference Vegetation Index (NDVI) was obtained from the ratio between the difference in the reflectivities of the near infrared (ρIV) and the red (ρV) and the sum between them (TUCKER, 1979 apud TASUMI, 2003):

$$(IVP - V) / (IVP + V) \qquad (3)$$

4.2.6 <u>Soil Adjusted Vegetation Index (SAVI)</u>

The Soil Adjusted Vegetation Index (SAVI) is an index that takes into account the effects of exposed soil in the images analysed, to adjust the NDVI when the surface is not completely covered by vegetation. SAVI (Equation 3) was developed by HUETE (1988) as a transformation technique to minimise the influence of soil reflectance on spectral vegetation indices involving the red and near-infrared wavelengths and to more accurately model near-infrared radiance in more open canopies (SILVA et al, 2011).

Although the NDVI index has been widely used, it has some limitations, which affect the results achieved, such as interference due to soil colour and moisture effects. An index was therefore developed that could improve NDVI values without the need for field measurements for each area analysed (JENSEN, 2009). To this end, an improved index was developed based on a constant, L, as an adjustment factor for the canopy substrate. Thus, the Soil Adjusted Vegetation Index (SAVI) is expressed as Equation 4:

$$SAVI = \frac{(1+L)(\rho IV - \rho V)}{(L + \rho IV + \rho V} \qquad (4)$$

Where: L is a constant known as the SAVI index adjustment factor, and can take values from 0.25 to 1 depending on the soil cover. According to Huete (1988) a value for L of 0.25 is indicated for dense vegetation and 0.5 for vegetation with intermediate density, when the value of L is 1 for vegetation with low density. If the SAVI value is equal to 0, its values become equal to the NDVI values. Therefore, the most commonly used L value is 0.5.

The constant L can have values from 0 to 1, varying according to the biomass itself. According to Huete (1988) apud Ponzoni and Shimabukuro (2009), the optimum values for L are: L = 1 (for low vegetation densities) L = 0.5 (for medium vegetation densities) L = 0.25 (for high vegetation densities) According to the aforementioned authors, in general the L = 0.5 factor is more commonly used, since it encompasses a greater variation in vegetation conditions. Even so, SAVI is limited in terms of the different biomes and agricultural situations, since the constant values are generalised and do not take into account the specific characteristics of the environments analysed, but only the vegetation densityPONZONI, SHIMABUKURO, 2009).

4.2.7 Leaf area index (LAI)

According to Pereira and Machado (1987) leaf area is a factor that depends on the number and size of leaves and the phenological stage. In order to estimate productivity and evapotranspiration from forest-soil-atmosphere interface models, IAF has been considered the main descriptor variable of the plant canopy in numerous studies (XAVIER et al, 2002). The calculation of IAF, which represents the ratio between the total area of all the leaves contained in a given pixel and the area of the pixel, was done using an empirical equation obtained by Allen et al. (2002):

$$IAF = -(\ln(0{,}69 - IVAS/0{,}59)/0{,}91) \qquad (5)$$

CHAPTER 5

RESULTS AND DISCUSSION

Firstly, the NDVI, which is related to the chlorophyll content of the vegetation, was calculated for the area around the Paratibe River. It can be seen that over six years the urban area has become denser and the Janga Forest area lost vegetative vigour from 2005 to 2011. Figure 6 - NDVI characterisation.

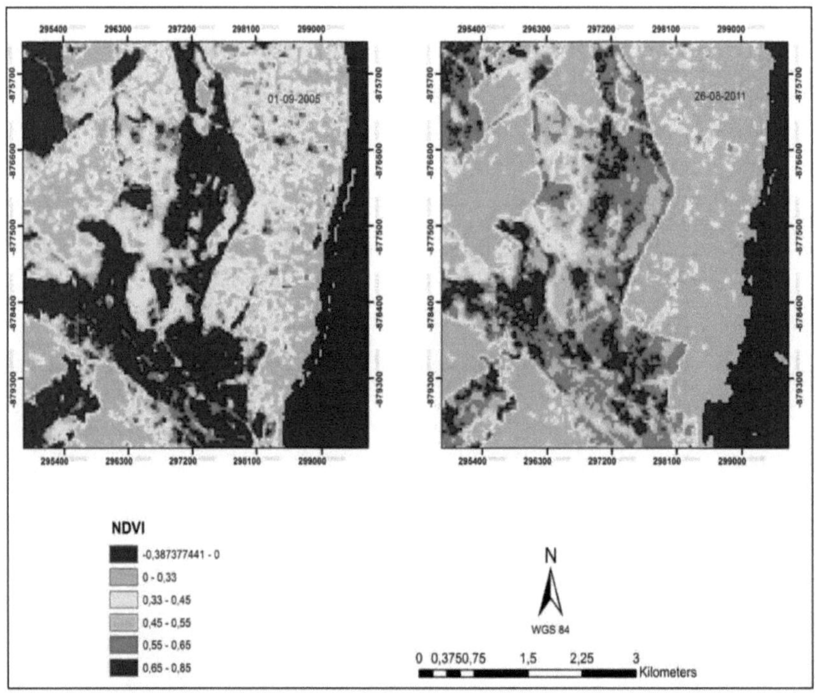

Source: Prepared by the author.

The NDVI (Figure 6) showed preserved vegetation until 2005, which is corroborated by the Socio-Environmental Diagnosis of the North Coast of Pernambuco (2001), which classified the Mata do Janga Ecological Reserve as being in a regular state of conservation, due to the occurrence of degraded stretches in the interior or on the edge as a result of excessive wood removal and rubbish dumping.

The index was higher than 0.65 in most areas around the forest, but in the centre of the forest there was an NDVI range of 0.33-0.45 in 2005. In 2011, in the same area, in the centre of the forest, a value between 0.45-0.55 was found, which indicates that the area has undergone a process of recovery and vegetation densification. We can also see that there has been a process of intensification of the urban area around the forest, where the predominant values around the area are between 0-0.33.

In 2011, the predominant NDVI in the Mata do Janga area was no longer above 0.65 and most of the pixels in the area showed values around 0.55-0.65, which means that the area has been under some stress, which may be related to the increase in urban densification that has occurred in the area around Mata do Janga (figure 6).

The next step in this work was to estimate SAVI. This index serves as the basis for estimating IAF. The map below shows how the vegetation around the Paratibe River has changed in just 6 years of sampling, figure 7.

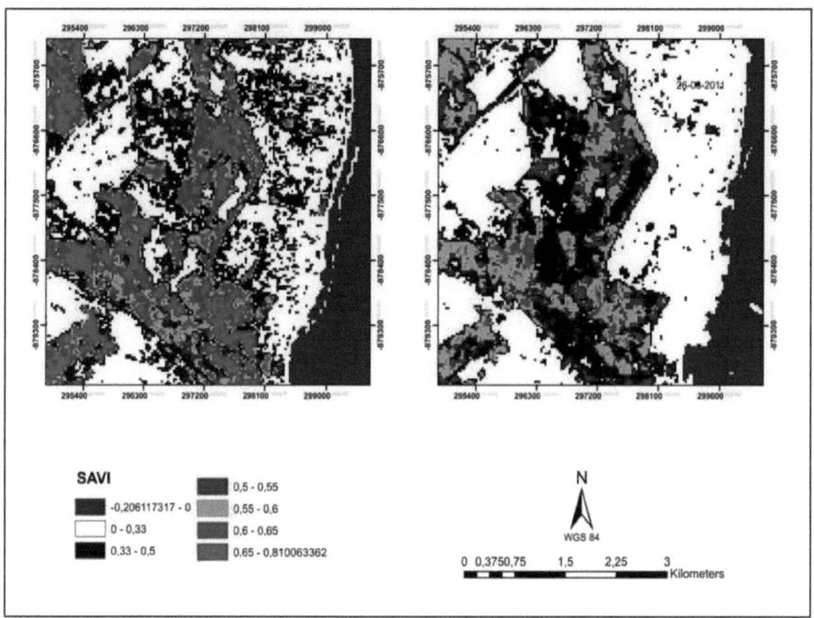

Figura 7 - Analysing the vegetation index at ground level.

Source: Prepared by the author.

As can be seen, in 2005 there were degraded areas within Mata do Janga (values between 0 and 0.33). These areas of exposed soil decreased in 2011, but the overall SAVI values decreased (Figure 7). In 2011, SAVI values of 0.33 to 0.5 predominated in the area, while in 2005 SAVI values were very high (greater than 0.65). In 2011, the

31

predominant values in the forest ranged from 0.33 to 0.6. In 2005, the predominant values were higher than 0.55 (figure 7).

The next image shows the IAF, which is related to the biomass content present in the area, i.e. it reflects the dynamics of the forest canopy and has to do with the leafiness of the vegetation (Figure 8).

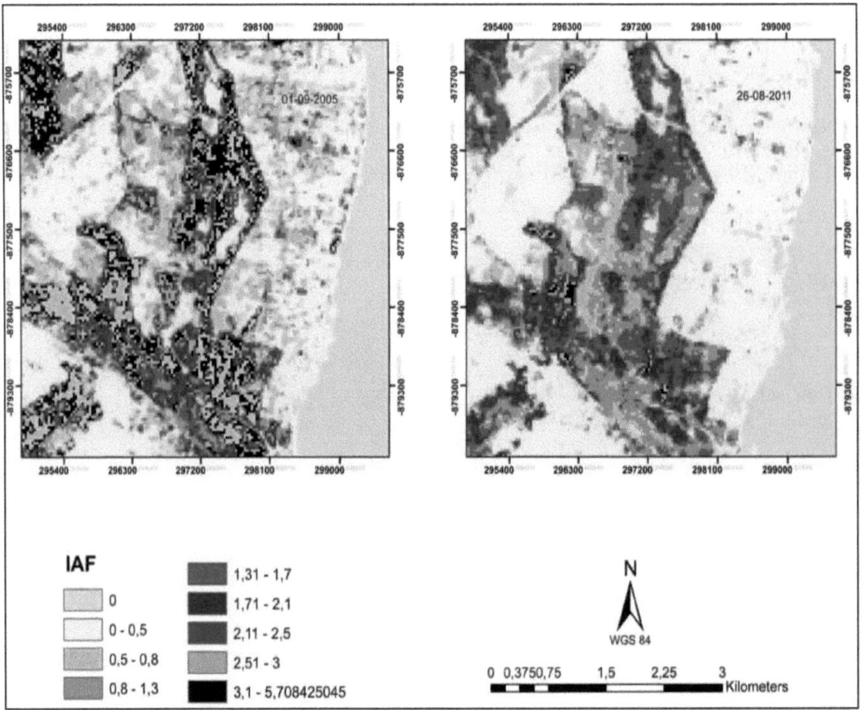

Figura 8   - Comparison of IAF.

Source: Prepared by the author.

It can be seen from the LAI results that the urban area/exposed soil showed rapid growth from 2005 to 2011 (0-0.5). In 2005, the LAI, despite showing lower values (0.5-0.8), also showed a greater amount of green areas than in 2011 (figure 8). The areas that resisted the urbanisation process were precisely those protected by environmental legislation, such as Mata do Janga. For this reason, the Leaf Area Index (LAI) is a biophysical variable that expresses the growth rate of a given plant community and is closely related to its productivity (LANG and MCMURTRIE, 1992; XAVIER et al, 2004; PAIVA et al, 2009).

It can also be seen that in some areas of the forest the 2011 IAF showed a recovery (oscillating between 1.31 and 2.1) in relation to the 2005 IAF (with predominant values of 0.5 to 1.71). As we've seen, the predominant NDVI in the Janga Forest decreased from 2005 to 2011, but the 2011 IAF values showed an increase.

This is because a high chlorophyll content in the vegetation does not always mean a high biomass content (often agricultural areas, depending on the successional stage of the vegetation, have even higher NDVI values than forest areas, as NDVI is influenced by the chlorophyll content in the vegetation). It is important to emphasise that IAF is considered to be the most important biophysical variable directly related to evapotranspiration (LANG AND MCMURTRIE, 1992; SELLERS et al., 1997; XAVIER AND VETTORAZZI, 2004) and to the canopy's ability to intercept rain (KERGOAT, 1998; DIJK AND BRUIJNZEEL, 2001).

It is also true that the 2005 IAF reached higher peaks than the 2011 image from the Landsat 5 satellite (showing values greater than 2.51 to peaks greater than 5), while in 2011 these peaks greater than 2.51 were rare.

CHAPTER 6

FINAL CONSIDERATIONS

The delimitation of Mata do Janga as an environmental protection area has shown effective results in the conservation of the area, since until 2005 the vegetation indices used showed that there were pockets of degradation within the forest. However, the structure of the vegetation has not increased over the years, which indicates that despite the delimitation of the protected area of the forest, there is still environmental stress in the area.

The index also shows that there was greater urban densification between 2005 and 2011, which led to the loss of green areas in the municipality of Paulista. This is a risk factor for the protected areas around the Paratibe River.

BIBLIOGRAPHICAL REFERENCES

ANDRADE-LIMA, D. **Estudos fitogeográficos de Pernambuco.** Arquivo do Instituto de Pesquisas Agronômicas de Pernambuco, Recife, 5: 305-341. 1960.

AVERY, T. E.; BERLIN, G. L. **Fundamentals of remote sensing and airphoto interpretation.** 5. ed. New Jersey: Prentice Hall, 1992.

BARBOSA, K. M. do N. **Spatial monitoring of biomass and organic carbon of floodplain herbaceous vegetation in Central Amazonia.** Curitiba: Federal University of Paraná, Doctoral Thesis, 131 p., 2006.

BORATTO, I. M. de P.; GOMIDE, R. L. **Application of the NDVI, SAVI and IAF vegetation indices to characterise vegetation cover in the northern region of Minas Gerais.** Anais XVI Simpósio Brasileiro de Sensoriamento Remoto - SBSR, Foz do Iguaçu, PR, Brazil, INPE. 13 to 18 April 2013.

BRAZ, E. C. F.; Fábio AMORIM, J. de; GUERRA, T. N. F. **The Atlantic Forest inserted in the didactic contents and paradidactic activities of a school located in a state conservation unit - PE.** João Pessoa, p 850. October 2011.

CAMPANILI, M. II. PROCHNOW, M. **Mata Atlântica** - a network for the forest. Organisers Maura Campanili and Miriam Prochnow Brasília: RMA, 332p.: ill.; 30cm ISBN: 85-99824-01-5. 2006

CHANDER, G. MARKHAM, B. Revised Landsat-5 TM Radiometric Calibration Procedures and Postcalibration Dynamic Ranges. **IEEE Transactions on Geoscience and Remote Sensing**, v, 41, n. 11, 2003.

CLOUGH, B. F. **Primary productivity and growth of mangrove forests**. In: Robertson, A. I.; Alongi, D. M. (ed.). Tropical mangrove ecosystems. Coastal and estuarine studies #41. Washington: American Geophysical Union, p.225-250, 1992.

CONAMA. **Resolution No. 4 of 18 September 1985.** "Provides for definitions and concepts of Ecological Reserves". Published in the Official Gazette on 20 January 1986, pages 1095-1096. 1985.

CORRÊA, F. **A Reserva da Biosfera da Mata Atlântica: roteiro para o entendimento de seus objetivos e seu sistema de gestão**. São Paulo: National Council of the Atlantic Forest Biosphere Reserve, 1995.

COSTA J.E.R.. **Beach morphodynamics in the municipality of Paulista-PE**. Specialisation Monograph in Oceanography, Department of Oceanography, Federal University of Pernambuco. 2002

COSTA-LIMA, M. L. F. da. **The Atlantic Forest Biosphere Reserve in Pernambuco:** Current situation, actions and prospects . Cadernos da Reserva da Biosfera da Mata Atlântica Series. São Paulo, 1998.

CPRH. PERNAMBUCO ENVIRONMENT COMPANY. ENVIRONMENTAL **IMPACT REPORT** - RIMA. Implementation of structuring actions in the park sector of the Paratibe River urban and environmental conservation area (ZECUA) in the municipality of Paulista/PE October 2013

___ . PERNAMBUCO ENVIRONMENTAL COMPANY. **NORTH COAST SOCIO-ENVIRONMENTAL DIAGNOSIS**. Recife, 214p. 2003.

DIJK, A.I.J.M.; BRUIJNZEEL, L.A. Modelling rainfall interception by vegetation of variable density using an adapted analytical model. Part 2 - Model validation for a tropical upland mixed cropping system. **Journal of Hydrology**, Amsterdam, v.247, p.239-62, 2001.

EPIPHANIO, J. C. N.; GLERIANI, J. M.; FORMAGGIO, A. R.; RUDORFF, B. F. T. Vegetation indices in remote sensing of bean crops. **Pesquisa agropecuária brasileira**, Brasília, v. 31, n. 6, p. 445-454, 1996.

FALKENBERG, D. de B. **Aspectos da flora e da vegetação secundária da restinga de Santa Catarina, Sul do Brasil.** Insula, Florianópolis, n. 28, p. 1-30, 1999.

FISNER M. **Percepção dos Usuários sobre os efeitos das Obras de Proteção de Costa nas Praias de Casa Caiada (Olinda) e Janga (Paulista).** Master's Dissertation, Postgraduate Programme in Oceanography, Department of Oceanography, Federal University of Pernambuco, 133p. 2008.

GONZAGA DE CAMPOS. **Forest Map.** Rio de Janeiro: Geological and Mineralogical Service of Brazil, 1912.

KERGOAT, L. A model for hydrological equilibrium of leaf area index on a global scale. **Journal of Hydrology**, Amsterdam, v.212-13, p.268-86, 1998.

JENSEN, J. R. **Remote sensing of the environment:** a perspective on terrestrial resources / translated by José Carlos Neves Epiphanio et al. São José dos Campos. SP. 2009.

__________ . **Remote sensing of the environment:** an Earth Resources Perspective. 2. ed. Upper Saddle River: PrenticeHall. 592p. 2007.

LANG, A.R.G; MCMURTRIE, R.E. Total leaf areas of single trees of Eucalyptus grandis estimated from transmittances of the sun's beam. Agriculture! and Forest Meteorology, Amsterdam, v.58, p.79-92, 1992

MANSO V.A.V., COUTINHO P.N., GUERRA N.C., SOARES J.R. In: Pernambuco. Muehe D. (ed.) **Erosion and progradation of the Brazilian coast**. Ministério do Meio Ambiente Brasília, 179-196p Financiadora de Estudos e Projetos/Universidade Federal de Pernambuco - FINEP/UFPE. 2009. Integrated Environmental Monitoring - MAI-PE. Final Report - Vols. 1, 2 and 3, Recife, 485p. 2006b.

MANTOVANI, W. The degradation of Brazilian biomes. In: W.C. Ribeiro (ed.). **Brazilian environmental heritage**. pp. 367-439. Editora Universidade de São Paulo, São Paulo. 2003

MARKHAM, B. L.; BARKER, L. L. Thematic mapper bandpass solar exoatmospherical irradiances. International. **Journal of Remote Sensing**, v.8, n.3, p.517-523, 1987.

MENESES, P. R. **Fundamentals of Spectral Optical Radiometry**. In: MENESES, P. R. 2001.

MMA. **Atlantic Forest: environmental suitability manual** / Maura Campanili and Wigold Bertoldo Schaffer. - Brasília: MMA/SBF, 2010. 96 p. ; ill. colour. (Biodiversity Series, 35). 2010

MONTEIRO, J. J. F. **DEGRADATION OF THE MATA DO JANGA URBAN FOREST RESERVE (RESTINGA FOREST) - PAULISTA - PE**. Candido Mendes University. Specialisation Monograph. 2014.

MYERS, N. et al. Biodiversity hotspots for conservation priorities. **Nature**, v. 403, p.853- 858, 2000.

OTTO, H. **Ecosystems of the Atlantic Forest:** Restinga <URL: http://www.peruibest.com.br/colunas/meio-ambiente/ecossistemas-da-mataatlantica-

restinga.htm> date accessed 16-11-16.

PAIVA, Y.G.; RIBEIRO, A.; ALMEIDA, A.Q.; GLERIANE, J.M.; PEZZOPANE, J.E.M. Estimation of Leaf Area Index (LAI) using Hemispherical Photographs and Vegetation Indices in Eucalyptus clonal plantations. Anais... Proceedings of the **XIV Brazilian Symposium on Remote Sensing**, Natal, Brazil, 25-30 April, INPE, p. 2873-2880. 2009

PEREIRA, L. C. C. **Procesos litorales a largo las playas de Casa Caiada y Rio Doce, Olinda-PE (Brasil): implicaciones para gestión costera.** 276 f. Thesis (Doctorate in Marine Sciences) - Departament d'Enginyeria Hidráulica, Marítima i Ambiental, Universitat Politécnica de Catalunya, Barcelona. 2001.

PONZONI, F. J. Spectral Behaviour of Vegetation. In: MENESES, P. R., NETTO, J. S. M. (org) **Remote sensing, reflectance of natural targets.** Brasília - DF: Editora Universidade de Brasília - UNB, Embrapa Cerrados, p 157-199, 2001.

ATLANTIC FOREST BIOSPHERE RESERVE. Summary text: **The Atlantic Forest.** Available at:<< http://www.rbma.org.br/anuario/mata_01_sintese.asp>> Accessed on: 18 January 2017.

SELLERS, P.J.; DICKINSON, R.E.; RANDALL, D.A.; BETTS, A.K.; HALL, F.G.; BERRY, J.A.; COLLATZ, G.J.; DENNING, A.S.; MOONEY, H.A.; NOBRE, C.A.; SATO, N.; FIELD, C.B.; HENDERSON-SELLERS, A. **Modelling the exchanges of energy, water, and carbon between continents and the atmosphere. Science,** v.275, p.502-509, 1997

SILVA, E.R.A.C., GALVÍNCIO, J.D., BRANDÃO NETO, J.L., MORAIS, Y.C.B. Space-Time Analysis of Environmental Changes and their Reflection on the Development of Phenological of Vegetation of Mangrove. **Journal of Agriculture and Environmental Sciences** 4, 245-253. DOI: 10.15640/jaes.v4n1a30. 2015.

SILVA, M. N. DE A.; AUGUSTO COPQUE, C. DA S. M.; GIUDICE, D. S. **CONSEQUÊNCIAS DAS TRANSFORMAÇÕES AMBIENTAIS NO PROCESSO DE EXPANSÃO DAS CIDADES** - O EXEMPLO DE SALVADOR/BAHIA. UFBA, 2008.

SOUZA, M. M. A; SAMPAIO, E.V.S.B. Temporal variation in the structure of the mangrove forests of Suape-PE after the construction of the harbour. **Acta botânica brasílica**, v. 15, n.1, p. 112, 2001.

XAVIER, A.C., VETTORAZZI, C.A. **Monitoring leaf area index at watershed level Through NDVI from LANDSAT-7/ETM+ data.** Sci. Agric. (Piracicaba, Braz.), v.61, n.3, p.243-252, Anais XV Simpósio Brasileiro de Sensoriamento Remoto - SBSR, Curitiba, PR, Brazil, 30 April to 05 May 2011, INPE p.2112. May/June 2004.

XAVIER, A.C., VETTORAZZI, C.A., MACHADO, R.E. **Relationship between Leaf Area Index and pure component fractions of the Linear Spectral Mixture Model, using ETM+/LANDSAT images.** Eng. Agríc., Jaboticabal, v.24, n.2, p.421-430, May/Aug. 2004

WATSON, D.J. **Comparative physiological studies on growth of fields crops.** I - Variation in net assimilationrate and leaf area between species and varieties, and within and between years. Annals of Botany, London, v.11, p.41-76. 1947.

yes
I want morebooks!

Buy your books fast and straightforward online - at one of world's fastest growing online book stores! Environmentally sound due to Print-on-Demand technologies.

Buy your books online at
www.morebooks.shop

Kaufen Sie Ihre Bücher schnell und unkompliziert online – auf einer der am schnellsten wachsenden Buchhandelsplattformen weltweit! Dank Print-On-Demand umwelt- und ressourcenschonend produzi ert.

Bücher schneller online kaufen
www.morebooks.shop

Printed by Books on Demand GmbH, Norderstedt / Germany